THE MIND OF MACHINES

Exploring the Visionary World of Geoffrey Hinton's Artificial Intelligence

By

Michael J. Town

TABLE OF CONTENTS

Introduction

Artificial intelligence AI is revolutionizing the world in ways that have never been seen before. From virtual assistants to autonomous vehicles, AI is making our lives easier, more efficient, and more connected. But what is AI? How did it evolve? And who are the visionaries behind it? In this book, we explore the visionary world of Geoffrey Hinton's artificial intelligence.

The Evolution of Artificial Intelligence

The roots of artificial intelligence can be traced back to the early 20th century when researchers began developing machines that could perform simple tasks. Over the decades, AI has evolved to include machine learning, deep learning, and neural networks, among other technologies. Today, AI is used in a wide range of applications, from speech recognition to natural language processing to image recognition.

Geoffrey Hinton's Contributions to AI

Geoffrey Hinton is an esteemed figure in the field of computer science and a trailblazer in the development of deep learning techniques. He has made significant contributions to the field of artificial intelligence, particularly in the development of neural networks. Hinton's work has helped to push the boundaries of AI, making it possible for machines to learn and reason like humans.

Overview of the Book's Structure

The book is divided into eight chapters, each of which explores a different aspect of Geoffrey Hinton's artificial intelligence. Here's a brief overview of what you can expect in each chapter:

Chapter 1: Geoffrey Hinton's Early Life and Work

In this chapter, we learn about Geoffrey Hinton's early life and how he became interested in computer science. We also explore his early work in AI and the challenges he faced along the way.

Chapter 2: Convolutional Neural Networks

Convolutional neural networks are a type of neural network that is commonly used in image recognition. In this chapter, we explore how these networks work and how Hinton's work has contributed to their development.

Chapter 3: Recurrent Neural Networks

Recurrent neural networks are a type of neural network that is used in speech recognition and natural language processing. In this chapter, we explore how these networks work and how Hinton's work has contributed to their development.

Chapter 4: Deep Learning

Deep learning is a branch of machine learning that employs neural networks consisting of numerous layers. In this chapter, we explore how deep learning works and how Hinton's work has contributed to its development.

Chapter 5: Unsupervised Learning

Unsupervised learning is a type of machine learning where the algorithm learns from data without any labels. In this chapter, we explore how unsupervised learning works and how Hinton's work has contributed to its development.

Chapter 6: Cognitive Science and Neuroscience

In this chapter, we explore the relationship between AI and cognitive science and neuroscience. We also explore how Hinton's work has been influenced by these fields.

Chapter 7: Ethics and Society

AI raises many ethical and societal issues, including bias, privacy, and job displacement. In this chapter, we explore these issues and how Hinton's work addresses them.

Chapter 8: Future of Artificial Intelligence

In this chapter, we explore the future of AI and how Hinton's work is shaping it. We also discuss the potential benefits and challenges of AI in the coming years.

Conclusion

In the final part, we summarize the key takeaways from the book and reflect on the impact of Geoffrey Hinton's contributions to AI.

The Mind of Machines: Exploring the Visionary World of Geoffrey Hinton's Artificial Intelligence is a fascinating journey into the cutting-edge world of AI. Whether you're a researcher, a student, or simply curious about the future of technology, this book is sure to engage and inspire you.

Chapter One
Geoffrey Hinton's Early Life and Work

Geoffrey Hinton, one of the pioneers of artificial intelligence, was born in London, England, in 1947. He was born to a father who pursued mathematics and a mother who specialized in linguistics. Growing up, Hinton was exposed to both mathematics and language, which would later become the two driving forces behind his research.

Hinton received his undergraduate degree in experimental psychology from the University of Edinburgh in 1970. He proceeded to pursue a doctoral degree thereafter in artificial intelligence at the University of Edinburgh, which he completed in 1978. During his time at Edinburgh, Hinton became interested in neural networks, which were just beginning to gain traction as a field of research.

Early Work in Neural Networks

After completing his Ph.D., Hinton moved to the United States to work as a postdoctoral researcher at the University of Sussex. It was here that he began to develop his ideas about neural networks, and he quickly became known as a leading expert in the field.

In the early 1980s, Hinton collaborated with David Rumelhart and James McClelland on a series of papers that would have a significant impact on the field of artificial intelligence. These papers introduced the concept of distributed representations, which allows information to be processed in parallel across many different nodes in a network. This was a significant departure from the traditional rule-based systems that were prevalent at the time.

Collaboration with David Rumelhart and James McClelland

Hinton's collaboration with Rumelhart and McClelland was highly productive. They published a series of papers that laid the foundation for much of the work that followed in the field of artificial intelligence. In particular, they introduced the concept of backpropagation, which is an algorithm for training neural networks. Backpropagation works by adjusting the weights of the connections between nodes in a network, based on the error in the output of the network. This allows the network to learn to produce more accurate outputs over time.

Introduction of Backpropagation Algorithm
The introduction of the backpropagation algorithm was a significant breakthrough in the field of neural networks. It allowed networks to be trained more efficiently and effectively than ever before. This, in turn, led to the development of more complex networks that could perform a wide range of tasks, such as image and speech recognition.

Hinton continued to work on neural networks throughout the 1990s and 2000s, developing new algorithms and architectures that pushed the boundaries of what was possible with artificial intelligence. In 2012, he co-founded a company called DNN Research, which was later acquired by Google. At Google, Hinton continued to work on neural networks, and he was instrumental in the development of the Google Brain project, which aimed to develop a more powerful artificial intelligence system.

In conclusion, Geoffrey Hinton's early life and work set the stage for his groundbreaking contributions to the field of artificial intelligence. His work on neural networks and the backpropagation algorithm revolutionized the field, allowing for the development of more powerful and complex systems. Hinton's continued work in the field has led to significant advances in artificial intelligence, and his legacy will continue to shape the field for years to come.

Chapter Two
Convolutional Neural Networks

The development of convolutional neural networks (CNNs) has been one of the most significant breakthroughs in the field of artificial intelligence, particularly in computer vision. CNNs were first introduced in the 1980s by Yann LeCun and his colleagues, who were looking for a way to process handwritten digits for use in postal sorting machines. Since then, CNNs have undergone numerous modifications and advancements, making them the backbone of modern computer vision and image processing systems.

This chapter will explore the history of CNNs and their evolution, as well as their applications in computer vision. We will also examine the successes of CNNs in image recognition and classification and their current and future applications.

History and Evolution of Convolutional Neural Networks

The first CNNs were inspired by the visual system of animals, which uses a series of filters to process information from the environment. LeCun and his colleagues designed a CNN that used convolutional layers to extract features from handwritten digits, and a fully connected layer to classify them. This system, called LeNet-5, achieved state-of-the-art performance on the MNIST dataset and laid the groundwork for future developments in CNNs.

Over the years, researchers have made significant advancements in CNNs, including the development of deeper networks, the use of residual connections, and the implementation of various regularization techniques. One of the most significant breakthroughs came in 2012 when a team led by Alex Krizhevsky used a deep CNN to win the ImageNet Large Scale Visual Recognition Challenge (ILSVRC), a competition that required systems to classify over a million images into 1,000 different categories. This system, called AlexNet, achieved a top-5 error rate of 15.3%, significantly outperforming all other competitors.

Since then, CNNs have become the go-to method for image classification, with deeper and more complex networks achieving even better performance. For example, in 2015, a team from Microsoft Research developed a 152-layer CNN called ResNet, which achieved a top-5 error rate of 3.57% on the ImageNet dataset, surpassing human performance.

Applications in Computer Vision

CNNs have numerous applications in computer vision, including image classification, object detection, segmentation, and recognition. In image classification, CNNs take an input image and output a probability distribution over a set of predefined classes. Object detection involves localizing and classifying objects within an image, while segmentation involves labeling each pixel in an image with a class label.

CNNs are also used in facial recognition systems, self-driving cars, and medical image analysis, among other applications. For example, CNNs have been used to automatically detect cancer in medical images, which could help improve diagnosis and treatment.

Successes in Image Recognition and Classification

CNNs have achieved unprecedented success in image recognition and classification, outperforming traditional machine learning algorithms and even human performance in some cases. For example, in the 2015 ILSVRC competition, the top-performing CNN achieved a top-5 error rate of 3.57%, compared to a human error rate of 5%.

CNNs have also been used to create generative models that can produce realistic images, such as DeepDream and GANs (generative adversarial networks). These models have potential applications in art, design, and entertainment.

Current and Future Applications of CNNs

CNNs have numerous current and future applications, including improving image and video search, creating more realistic virtual and augmented reality experiences, and enhancing medical diagnosis and treatment. CNNs are also being used in robotics, natural language processing, and speech recognition.

One potential future application of CNNs is in autonomous vehicles, where they could help the vehicles identify and respond to objects in real time. CNNs could also be used to improve object detection in satellite imagery and to monitor wildlife populations.

Another promising area of research is the use of CNNs for personalized medicine. By analyzing medical images and patient data, CNNs could help predict disease risk and personalize treatment plans.

In addition, CNNs could be used to improve cybersecurity by identifying and blocking malicious images and videos or to enhance the accuracy and efficiency of search engines.

In conclusion, convolutional neural networks have revolutionized the field of computer vision and image processing, achieving unprecedented success in image recognition and classification. With their numerous applications in fields such as medicine, robotics, and cybersecurity, CNNs have the potential to transform many aspects of our daily lives.

As research in this field continues to evolve and new developments emerge, the possibilities for the future of CNNs are truly limitless. The mind of machines, powered by CNNs and other forms of artificial intelligence, will continue to inspire and transform the world we live in.

Chapter Three
Recurrent Neural Networks

Recurrent neural networks (RNNs) have emerged as one of the most exciting and powerful developments in artificial intelligence in recent years. These neural networks are uniquely suited for processing sequential data and have found widespread use in natural language processing, speech recognition, and image analysis, among other applications.

In this chapter, we will explore the development of RNNs, their applications in natural language processing, their successes in language modeling and machine translation, and their current and future applications.

Development of Recurrent Neural Networks

The development of RNNs can be traced back to the 1980s when they were first proposed as a way to model time-series data. However, early RNNs suffered from the "vanishing gradient" problem, which made it difficult to train them effectively.

In the 1990s, a breakthrough was made with the development of the Long Short-Term Memory (LSTM) architecture by Sepp Hochreiter and Jürgen Schmidhuber. LSTMs are a type of RNN that are designed to overcome the vanishing gradient problem by incorporating "memory cells" that can store information over long periods.

Since then, there have been numerous advances in the development of RNNs, including the development of Gated Recurrent Units (GRUs) and the use of attention mechanisms to improve their performance.

Applications in Natural Language Processing

RNNs have found widespread use in natural language processing, where they are used for tasks such as language modeling, sentiment analysis, and machine translation.

Language modeling involves predicting the likelihood of a sequence of words given some context. RNNs are particularly well-suited for this task, as they can take into account the entire history of the sequence.

Sentiment analysis is the task of identifying the emotional character or attitude conveyed by a written piece of text. RNNs can be trained to classify text into positive or negative sentiment, and have achieved state-of-the-art performance on benchmark datasets.

Machine translation refers to the process of converting written text from one language into another. RNNs have been used for this task with great success, with models such as the Transformer achieving state-of-the-art performance.

Successes in Language Modeling and Machine Translation

One of the key successes of RNNs in natural language processing has been in the area of language modeling. In 2010, a team led by Yoshua Bengio achieved a breakthrough in language modeling by training a neural network with multiple layers of non-linear processing units. This model, known as a deep neural network, was able to outperform previous state-of-the-art models on benchmark datasets.

Since then, RNNs have continued to push the boundaries of language modeling, with models such as the LSTM and the Transformer achieving even better performance.

RNNs have also achieved great success in machine translation. In 2016, the Google Brain team introduced the Google Neural Machine Translation (GNMT) system, which used an attention mechanism and an encoder-decoder architecture with LSTMs. This system achieved state-of-the-art performance on several benchmark datasets and has since become the basis for Google Translate.

Current and Future Applications of RNNs

RNNs are being used in an increasing number of applications, from predicting stock prices to generating music. In the field of natural language processing, RNNs are being used for tasks such as text summarization, question answering, and dialogue generation.

One promising application of RNNs is in the field of speech recognition. RNNs have been used to improve the accuracy of speech recognition systems by modeling the temporal dependencies in speech signals.

Another promising application of RNNs is in the field of robotics. RNNs can be used to learn the dynamics of a robot's environment, allowing it to predict future states and plan actions accordingly.

RNNs can also be used in the field of computer vision, where they have been used for tasks such as image captioning and video analysis. In image captioning, RNNs can be used to generate natural language descriptions of images, while in video analysis, they can be used to predict future frames in a video sequence.

In the future, RNNs are likely to continue to play a key role in advancing the field of artificial intelligence. With the development of more powerful hardware and the continued refinement of

RNN architectures, RNNs will likely become even more effective at processing sequential data and performing complex tasks.

Geoffrey Hinton's contributions to RNN research
Geoffrey Hinton is a pioneer in the field of AI and has made significant contributions to the development of RNNs. He is best known for his work on deep learning and his contributions to the development of the backpropagation algorithm.

Hinton has also made significant contributions to the development of RNNs. In particular, he has worked on developing new architectures such as the LSTM and the GRU, which have been instrumental in the success of RNNs in natural language processing.

Recurrent neural networks have emerged as a powerful tool for processing sequential data and have found widespread use in natural language processing, speech recognition, image analysis, and robotics. With continued advances in RNN architectures and the development of more powerful hardware, RNNs will likely continue to play a key role in advancing the field of artificial intelligence. The future of RNNs looks bright, and we can expect to see many exciting new applications in the years to come.

Chapter Four
Deep Learning

Deep learning is a subfield of machine learning that is inspired by the structure and function of the human brain. It involves the use of artificial neural networks that are capable of learning and improving upon tasks by processing large amounts of data. This chapter will delve into the work of Geoffrey Hinton, who is widely regarded as the "Godfather of Deep Learning," and the advancements in this field that have led to its widespread applications in speech recognition, object detection, and gaming. We will also discuss some of the criticisms and limitations of deep learning.

Hinton's Work in Deep Learning

Geoffrey Hinton is a computer scientist and cognitive psychologist who has made significant contributions to the field of deep learning. His work has focused on developing algorithms for artificial neural networks that can learn and improve upon tasks with minimal supervision. One of Hinton's most significant contributions to the field was the development of deep belief networks (DBNs) in the 1990s. DBNs are probabilistic models that are used to represent and learn hierarchical structures in data.

Hinton also played a key role in the development of backpropagation algorithms, which are used to train artificial neural networks by adjusting the weights of connections between neurons. This technique was crucial in the development of deep learning algorithms, as it allowed neural networks to learn complex representations of data by adjusting the weights of connections between layers.

In 2012, Hinton and his team won the ImageNet Large Scale Visual Recognition Challenge, which is an annual competition that aims to develop algorithms for object recognition in images. Their algorithm, known as AlexNet, achieved a significant improvement in object recognition accuracy compared to previous algorithms. This breakthrough led to a surge of interest in deep learning and paved the way for its widespread adoption.

Advancements in Deep Neural Networks

Since Hinton's early work on deep belief networks and backpropagation, there have been significant advancements in deep neural networks. One such advancement is the development of convolutional neural networks (CNNs), which are used in image and speech recognition tasks. CNNs are designed to process data with a grid-like topology, such as images, by applying convolution operations to identify patterns in the data.

Another advancement in deep neural networks is the development of recurrent neural networks (RNNs), which are used in tasks that involve sequential data, such as speech recognition and natural language processing. RNNs are designed to process data with a temporal dimension by maintaining a state that can be updated based on the current input and the previous state.

More recently, there has been a surge of interest in generative models, such as generative adversarial networks (GANs) and variational autoencoders (VAEs). These models are capable of generating new data samples that are similar to the training data, which has applications in areas such as image and music generation.

Applications in Speech Recognition, Object Detection, and Gaming

Deep learning has been applied to a wide range of applications, including speech recognition, object detection, and gaming. In speech recognition, deep neural networks are used to process audio data and convert it into text. This technology is used in virtual assistants, such as Siri and Alexa, and has also been applied to language translation.

In object detection, deep neural networks are used to identify objects in images and videos. This technology is used in self-driving cars to detect pedestrians, traffic signs, and other vehicles. It is also used in security cameras to detect intruders and in medical imaging to detect abnormalities.

In gaming, deep neural networks have been used to develop agents that can play games at a superhuman level. For example, AlphaGo, developed by DeepMind, was able to defeat the world champion in the game of Go, which was previously thought to be too complex for computers to play at a high level. This achievement was a significant breakthrough in the field of artificial intelligence and demonstrated the potential of deep learning in complex decision-making tasks.

Criticisms and Limitations of Deep Learning

Despite the many advancements and applications of deep learning, there are also criticisms and limitations to this technology. One criticism is the "black box" problem, which refers to the lack of interpretability of deep neural networks. The complex structure of deep neural networks makes it difficult to understand how they arrive at their decisions, which can be problematic in applications where explanations are required, such as in healthcare and finance.

Another criticism is the need for large amounts of labeled data to train deep neural networks. This requirement can be costly and time-consuming, especially in applications where labeled data is scarce or difficult to obtain.

There are also limitations to the generalizability of deep neural networks. In some cases, deep neural networks may perform well on the training data but fail to generalize to new, unseen data. This problem, known as overfitting, can be addressed through techniques such as regularization and data augmentation, but it remains a challenge in deep learning.

In this chapter, we have explored the work of Geoffrey Hinton and the advancements in deep neural networks that have led to their widespread applications in speech recognition, object detection, and gaming. We have also discussed some of the criticisms and limitations of deep learning, including the "black box" problem, the need for large amounts of labeled data, and the challenge of generalization. Despite these limitations, deep learning continues to be an exciting area of research and has the potential to revolutionize many industries in the years to come.

Chapter Five
Unsupervised Learning

In recent years, unsupervised learning has emerged as a promising area of research in artificial intelligence. This chapter explores the work of one of the pioneers in the field, Geoffrey Hinton, and the advancements in unsupervised learning, including autoencoders and generative models. Additionally, this chapter discusses the applications of unsupervised learning in data compression and image generation, as well as the challenges and future directions in the field.

Hinton's Research In Unsupervised Learning

Geoffrey Hinton is a renowned computer scientist who has made significant contributions to the field of unsupervised learning. His work on deep learning and neural networks has revolutionized the field of artificial intelligence. Hinton's research has focused on developing algorithms that can learn representations of data without explicit supervision.

One of Hinton's most significant contributions to unsupervised learning is the concept of restricted Boltzmann machines (RBMs). RBMs are a type of neural network that can learn the underlying structure of data without any labeled examples. These networks are trained to reconstruct input data, which forces them to learn the underlying features of the data.

Advancements In Autoencoders And Generative Models

Autoencoders and generative models are two areas of unsupervised learning that have seen significant advancements in recent years. Autoencoders are neural networks that can learn to compress data into a lower-dimensional representation and then reconstruct the original data from the compressed representation. They have been used in a variety of applications, such as image compression and data denoising.

Generative models, on the other hand, are networks that can learn to generate new data that is similar to the training data. These models have been used in a variety of applications, such as image and speech generation. One of the most popular generative models is the generative adversarial network (GAN), which consists of two networks: a generator and a discriminator. The generator's task is to acquire the ability to produce novel data, while the discriminator's objective is to become skilled at recognizing the difference between genuine and synthesized data.

Applications In Data Compression And Image Generation

There are numerous domains in which unsupervised learning finds utility, owing to its diverse range of applications. One of the most significant applications is data compression. Autoencoders can learn to compress data into a lower-dimensional representation, which can significantly reduce the storage space required to store the data.

Another significant application of unsupervised learning is image generation. Generative models, such as GANs, can learn to generate new images that are similar to the training data. These models have been used in a variety of applications, such as creating realistic images of people or animals.

Challenges And Future Directions In Unsupervised Learning

Despite the significant advancements in unsupervised learning, there are still several challenges that need to be addressed. One of the most significant challenges is the lack of interpretability of unsupervised models. Unlike supervised models, which can be easily interpreted, unsupervised models can be challenging to understand.

Another challenge is the lack of robustness of unsupervised models. These models can be sensitive to changes in the input data, which can lead to significant changes in the output. Additionally, unsupervised models can be challenging to train, as they often require a large amount of data and computational resources.

The future of unsupervised learning is promising, with many researchers working to address the challenges in the field. One area of research that has shown promise is the development of more interpretable unsupervised models. Additionally, researchers are working on developing more robust unsupervised models that can handle noisy or incomplete data.

Unsupervised learning has emerged as a promising area of research in artificial intelligence, with significant advancements in autoencoders and generative models. The work of Geoffrey Hinton has been instrumental in advancing the field of unsupervised learning. Hinton's research on restricted Boltzmann machines (RBMs) has provided a foundational framework for unsupervised learning, and his work on deep learning and neural networks has revolutionized the field of artificial intelligence.

Advancements in autoencoders and generative models have also significantly contributed to the development of unsupervised learning. Autoencoders have been used in a variety of applications, such as data compression and data denoising, while generative models have been used in image and speech generation.

Unsupervised learning has many potential applications, including data compression and image generation. However, there are still several challenges that need to be addressed, such as the lack of interpretability and robustness of unsupervised models.

Despite these challenges, the future of unsupervised learning is promising. Researchers are working on developing more interpretable and robust unsupervised models, as well as exploring new directions in the field. As unsupervised learning continues to advance, it has the potential to significantly impact many fields, from computer vision and natural language processing to robotics and autonomous systems.

Chapter Six
Cognitive Science and Neuroscience

Geoffrey Hinton's interest in cognitive science and neuroscience stems from his belief that artificial intelligence (AI) can only truly reach its potential when it is grounded in a deep understanding of how the human brain works. In this chapter, we will explore Hinton's contributions to the fields of cognition and perception, as well as the connections between AI and neuroscience. We will also examine the future directions for interdisciplinary research in these areas.

Contributions to Theories of Cognition and Perception

Hinton's contributions to the field of cognitive science and neuroscience are vast and varied. He has been instrumental in developing theories of cognition and perception that have revolutionized the way we think about AI and its relationship to the human mind. One of Hinton's most significant contributions has been his work on deep learning, a subfield of machine learning that involves training neural networks with multiple layers. This approach has allowed AI systems to achieve unprecedented levels of accuracy in tasks such as image and speech recognition.

Another area in which Hinton has made significant contributions is in the field of unsupervised learning. This approach involves training AI systems on large datasets without providing them with explicit labels or instructions. Instead, the systems must learn to identify patterns and structures in the data on their own. Hinton's work on unsupervised learning has led to the development of several groundbreaking AI algorithms, including the Deep Boltzmann Machine and the Variational Autoencoder.

Hinton has also been a leading figure in the field of probabilistic modeling, which involves developing mathematical models that can represent uncertain or incomplete information. This approach has been particularly useful in areas such as natural language processing, where the meaning of a sentence may depend on context and other factors. Hinton's work on probabilistic modeling has led to the development of several important AI algorithms, including the Restricted Boltzmann Machine and the Deep Belief Network.

Connections Between Ai And Neuroscience

Hinton's work in AI has been heavily influenced by his interest in neuroscience. He believes that by studying the brain, we can gain insights into how to build better AI systems. In particular, Hinton is interested in the way that the brain processes information and how this can be replicated in AI systems.

One area in which Hinton has drawn inspiration from neuroscience is in the development of convolutional neural networks (CNNs). These networks are designed to mimic the way that the visual cortex processes visual information. By using multiple layers of neurons that perform increasingly complex computations, CNNs can identify features in images and classify them with remarkable accuracy.

Another area in which Hinton has drawn inspiration from neuroscience is in the development of recurrent neural networks (RNNs). These networks are designed to mimic the way that the brain processes sequential information, such as speech or text. By using feedback loops that allow information to flow back into the network, RNNs can learn temporal dependencies in data and generate predictions based on this information.

Future Directions for Interdisciplinary Research

As AI and neuroscience continue to evolve, there are many exciting opportunities for interdisciplinary research. Hinton believes that by working together, researchers in these fields can make significant advances in understanding the human mind and building intelligent machines.

One area in which Hinton sees great potential is in the development of AI systems that are more flexible and adaptable. While current AI systems are very good at performing specific tasks, they struggle when faced with new or unexpected situations. By drawing on insights from neuroscience, Hinton believes that we can develop AI systems that are more robust and adaptable, able to learn from experience and adjust their behavior accordingly.

Another area in which Hinton sees great potential is in the development of AI systems that are more explainable. One of the biggest challenges facing AI today is the so-called "black box" problem. That is, while AI systems are often able to make accurate predictions or perform complex tasks, it can be difficult to understand how they arrived at these results. This lack of transparency makes it difficult to trust and debug AI systems, particularly in applications such as healthcare or finance where the stakes are high.

Hinton believes that by drawing on insights from cognitive science and neuroscience, we can develop AI systems that are more transparent and interpretable. For example, by designing AI algorithms that are more closely aligned with the way that the human brain processes information, we may be able to better understand how the algorithms are making decisions. This could lead to the development of more transparent and trustworthy AI systems, which could have a significant impact on a wide range of industries.

Finally, Hinton sees great potential in the development of AI systems that are more human-like in their behavior and interactions. While current AI systems can perform many tasks with remarkable accuracy, they lack the kind of common sense and intuition that humans possess. By drawing on insights from cognitive science and neuroscience, we may be able to develop AI systems that are more attuned to human needs and preferences, and that are better able to interact with humans in natural and intuitive ways.

Overall, the future of interdisciplinary research in cognitive science, neuroscience, and AI is incredibly promising. By working together, researchers in these fields can continue to make significant advances in our understanding of the human mind and the development of intelligent machines. Hinton's contributions to these fields have been significant, and his visionary approach to AI research continues to inspire and shape the work of researchers around the world.

Chapter Seven
Ethics and Society

As AI technology advances and becomes more widespread, it is important to consider the ethical implications of its development and deployment. In this chapter, we will explore the various considerations for the ethical use of AI, the views of AI pioneer Geoffrey Hinton on ethical AI, the societal implications of AI and automation, and the future responsibilities of AI researchers and practitioners.

Considerations for the Ethical Development and Deployment of AI

One of the most pressing ethical considerations for the development and deployment of AI is the potential for bias. AI algorithms are only as good as the data they are trained on, and if that data is biased, the resulting algorithm will also be biased. This can lead to discriminatory outcomes, such as the use of facial recognition technology that is more accurate on lighter-skinned individuals than on darker-skinned individuals.

Another consideration is transparency. As AI algorithms become more complex, it can be difficult for even the researchers who created them to understand exactly how they are making decisions. This lack of transparency can make it difficult to assess the fairness of an algorithm's decisions, and can also make it difficult to identify and correct bias.

Privacy is another important consideration. As AI technology becomes more advanced, it becomes easier to collect and analyze large amounts of data about individuals. This data can be used to make decisions about everything from job applications to credit scores. It is important to ensure that this data is collected and used ethically, with appropriate safeguards in place to protect individual privacy.

Finally, there is the question of accountability. As AI technology becomes more autonomous, it can be difficult to determine who is responsible for its decisions. This is particularly true in cases where an AI system causes harm. It is important to establish clear lines of accountability and responsibility to ensure that those who are harmed by AI systems can seek redress.

Hinton's Views on Ethical AI
Geoffrey Hinton is one of the most influential figures in the field of AI, and his views on ethical AI carry significant weight. Hinton has argued that it is important to build AI systems that are aligned with human values, and that can be trusted to make decisions that are in the best interests of society as a whole.

Hinton has also argued that transparency is key to ensuring the ethical use of AI. He believes that AI algorithms should be designed to be interpretable so that researchers and users can understand how they are making decisions. This transparency can help to identify and correct bias, and can also help to build trust in AI systems.

Societal Implications of AI and Automation

The widespread adoption of AI and automation is likely to have significant societal implications. The job market could be influenced as a result of this. As AI and automation take over more tasks that were previously done by humans, there is a risk that large numbers of workers could be displaced. This could lead to significant economic and social disruption.

Another potential impact is on privacy. As AI and automation become more advanced, it becomes easier to collect and analyze large amounts of data about individuals. This data can be used to make decisions about everything from job applications to credit scores. It is important to ensure that this data is collected and used ethically, with appropriate safeguards in place to protect individual privacy.

Finally, there is the question of accountability. As AI technology becomes more autonomous, it can be difficult to determine who is responsible for its decisions. This is particularly true in cases where an AI system causes harm. It is important to establish clear lines of accountability and responsibility to ensure that those who are harmed by AI systems can seek redress.

Future Implications and Responsibilities for AI Researchers and Practitioners

As AI technology continues to advance, it is likely to have an increasingly significant impact on society. This means that AI researchers and practitioners have a responsibility to ensure that their work is guided by ethical principles and is aimed at creating positive social outcomes. This includes being transparent about the development and deployment of AI, addressing bias and discrimination in AI systems, protecting individual privacy, and establishing clear lines of accountability and responsibility for the decisions made by AI systems.

Furthermore, AI researchers and practitioners must engage with a wide range of stakeholders, including policymakers, civil society organizations, and members of the public. This engagement is critical to ensuring that the development and deployment of AI are aligned with the needs and values of society as a whole.

In addition, AI researchers and practitioners must be mindful of the potential risks and unintended consequences of their work. This means taking a proactive approach to identifying and mitigating risks and being willing to revise or even abandon projects that could have negative impacts.

Ultimately, the future of AI depends on the responsible actions of AI researchers and practitioners. By prioritizing ethical considerations and working in collaboration with other stakeholders, they can help to ensure that AI technology is developed and deployed in a way that benefits society as a whole.

Chapter Eight
Future of Artificial Intelligence

In recent years, artificial intelligence (AI) has become one of the most rapidly developing fields of technology. With the advancement of machine learning and deep learning algorithms, AI is poised to transform many aspects of our lives in the coming years. In this chapter, we will explore current and future trends in AI research, predictions for advancements in AI technology, and the potential impact on society and the workforce. Additionally, we will delve into the visionary world of Geoffrey Hinton's artificial intelligence and his vision for the future of AI.

Current and Future Trends in AI Research

The field of AI research is constantly evolving, and new developments are emerging regularly. Currently, several trends in AI research are shaping the future of the field. The utilization of deep learning algorithms is becoming more prevalent and noteworthy as a prominent trend. These algorithms are designed to enable computers to learn from data, and they are highly effective in a wide range of applications, including image recognition, speech recognition, and natural language processing.

Another trend in AI research is the development of neural networks that are capable of simulating the behavior of the human brain. These neural networks are designed to learn from experience, and they can be used to recognize patterns in large datasets. Additionally, researchers are exploring the use of reinforcement learning algorithms, which enable machines to learn from feedback to improve their performance over time.

Predictions for Advancements in AI Technology

As AI technology continues to advance, there are several predictions for future advancements that are likely to occur. One of the most significant predictions is the development of more sophisticated natural language processing algorithms. These algorithms will enable machines to understand and interpret human language more accurately, which will have significant implications for fields such as healthcare and customer service.

Another prediction is the increasing use of AI in autonomous systems. This includes self-driving cars, drones, and other autonomous vehicles. As these systems become more advanced, they will be able to navigate complex environments with greater accuracy and reliability.

Finally, there is a growing interest in the development of AI systems that are capable of reasoning and decision-making. These systems will be able to process large amounts of data and make decisions based on that data, which will have significant implications for fields such as finance, healthcare, and law enforcement.

Potential Impact on Society and the Workforce

The rapid development of AI technology has the potential to have a significant impact on society and the workforce. On the one hand, AI has the potential to transform many industries, making them more efficient and productive. An instance of the application of AI systems is to carry out repetitive tasks automatically, thereby allowing workers to direct their attention towards intricate and imaginative work.

However, there are also concerns that the widespread adoption of AI could lead to job displacement and a widening income gap between those who have the skills to work with AI systems and those who do not. Additionally, there are concerns about the ethical implications of AI, including issues related to privacy, bias, and the potential misuse of AI systems.

Hinton's Vision for the Future of AI

Geoffrey Hinton is one of the most prominent researchers in the field of AI, and he has a unique vision for the future of the field. Hinton believes that the key to developing more advanced AI systems lies in the development of more sophisticated neural networks. Specifically, he believes that researchers need to focus on developing networks that are capable of hierarchical representation learning.

In this type of learning, networks can learn multiple levels of abstraction, which enables them to recognize complex patterns in data. Hinton believes that this type of learning is essential for developing AI systems that are capable of true human-level intelligence.

Additionally, Hinton believes that AI systems will eventually be able to learn from their own experiences, much like humans do. This will enable them to become more autonomous and adapt to new situations in real time. According to Hinton, this will be a significant breakthrough in the field of AI and will pave the way for the development of more advanced AI systems that can solve complex problems and make decisions in real-world scenarios.

Hinton also believes that the future of AI will involve more collaboration between humans and machines. He envisions a future where machines and humans work together to solve complex problems, with machines providing computational power and humans providing creativity and critical thinking skills.

Overall, Hinton's vision for the future of AI is one where machines are capable of true human-level intelligence, and where humans and machines work together to solve the most pressing challenges facing society.

In conclusion, the future of AI is an exciting and rapidly evolving field. As AI technology continues to advance, we can expect to see significant developments in areas such as natural language processing, autonomous systems, and reasoning and decision-making. However, there are also concerns about the potential impact of AI on society and the workforce, and it will be important to address these concerns as the technology continues to develop.

Through the visionary work of researchers such as Geoffrey Hinton, we can gain insight into the future of AI and the potential it holds for transforming many aspects of our lives. As we move forward, it will be important to continue to invest in AI research and development, while also ensuring that the technology is developed in a way that benefits society as a whole.

Conclusion

Geoffrey Hinton is one of the most influential figures in the field of artificial intelligence (AI) today. Throughout his career, he has made significant contributions to the development of neural networks and deep learning algorithms that have revolutionized the field of AI.

In this book, we have explored Hinton's visionary world of artificial intelligence, examining the key ideas and innovations that have shaped his work and exploring the implications of his research for the future of AI.

Summary of Hinton's Contributions to AI

Hinton's contributions to AI are many and varied, but perhaps his most significant contribution has been in the development of deep learning algorithms. Hinton was one of the pioneers of deep learning, and his work in this area has helped to transform the field of AI from a rule-based approach to a data-driven approach.

Hinton's work on neural networks has also been critical to the development of AI. Neural networks are computer systems modeled on the structure of the human brain, and they are designed to recognize patterns in data. Hinton's research has shown how neural networks can be used to achieve remarkable results in areas such as image and speech recognition.

Reflections on the Importance of Hinton's Work

The importance of Hinton's work cannot be overstated. His contributions to the field of AI have transformed the way we think about and approach the development of intelligent machines. Hinton's work has helped to move AI from a narrow set of applications to a wide range of real-world problems, from self-driving cars to medical diagnosis and drug discovery.

Hinton's work has also had a significant impact on the broader field of computer science. His contributions to deep learning have inspired many researchers to explore new ideas and techniques, and his work has helped to establish deep learning as one of the most promising areas of research in AI today.

Future Directions for AI Research and Development

As we look to the future, there are many exciting opportunities and challenges facing the field of AI. The development of more powerful and efficient computing systems is opening up new possibilities for AI research, and the growing availability of data is creating new opportunities for machine learning algorithms.

There are also significant challenges facing the field of AI, including concerns about the impact of AI on society and the workforce, as well as ethical and regulatory issues. Addressing these challenges will be critical to the responsible development and deployment of AI technologies in the years to come.

Geoffrey Hinton's work in the field of AI has been visionary and transformative. His contributions have helped to shape the way we think about and approach the development of intelligent machines, and his work will continue to inspire and guide researchers for years to come. As we look to the future, the challenges and opportunities facing the field of AI are immense, but with the visionary work of researchers like Hinton, we can be confident that the future of AI is bright.

www.ingramcontent.com/pod-product-compliance
Lightning Source LLC
Chambersburg PA
CBHW070854220526
45466CB00005B/1994